LIFE WITHOUT ANIMALS

WHAT IF BEES DISAPPEARED?

ANTHONY ARDELY

Gareth Stevens PUBLISHING

Please visit our website, www.garethstevens.com. For a free color catalog of all our high-quality books, call toll free 1-800-542-2595 or fax 1-877-542-2596.

Library of Congress Cataloging-in-Publication Data

Names: Ardely, Anthony, author.
Title: What if bees disappeared? / Anthony Ardely.
Description: New York : Gareth Stevens Publishing, [2023] | Series: Life without animals | Includes index.
Identifiers: LCCN 2021043910 | ISBN 9781538276242 (set) | ISBN 9781538276259 (library binding) | ISBN 9781538276235 (paperback) | ISBN 9781538276266 (ebook)
Subjects: LCSH: Bees–Juvenile literature. | Bees–Behavior–Juvenile literature. | Bees–Conservation–Juvenile literature.
Classification: LCC QL565.2 .A73 2023 | DDC 595.79/9–dc23
LC record available at https://lccn.loc.gov/2021043910

Published in 2023 by
Gareth Stevens Publishing
111 East 14th Street, Suite 349
New York, NY 10003

Designer: Rachel Rising
Editor: Kate Mikoley

Photo credits: Cover, p. 1 Suti Stock Photo/Shutterstock.com; pp. 3, 4, 6, 8, 10, 12, 14, 16, 18, 20, 22-24 Igor Kyrlytsya/Shutterstock.com; p. 5 szefei/Shuttestock.com; p. 7 Dennis van de Water/Shutterstock.com; p. 7 Daniel Prudek/Shutterstock.com; p. 9 BigBlueStudio/Shutterstock.com; p. 11 baismartin/Shutterstock.com; p. 13 RUKSUTAKARN studio/Shutterstock.com; p. 15 Cristian Storto/Shutterstock.com; p. 17 Mazur Travel/Shutterstock.com; p. 19 MakroBetz/Shutterstock.com; p. 21 ASakoulis/Shutterstock.com.

Printed in the United States of America

CPSIA compliance information: Batch #CSGS23: For further information contact Gareth Stevens, New York, New York at 1-800-542-2595.

CONTENTS

Boldface words appear in the glossary.

We Need Bees!

If you've ever been stung by a bee, you may have wished these **insects** would just disappear. Bees might seem like pests, but they are actually very important animals. Without them, we wouldn't have many of the foods we eat!

Bees can be found in every part of the world except Antarctica. More than 20,000 species, or kinds, **exist**. The most well-known are bumblebees and honeybees. Bees are often hairy. Many are black and have yellow or brown markings.

BUMBLEBEE

HONEYBEE

Hungry Bees

If you've ever enjoyed honey, you can thank honeybees! Bees get their food from the nectar and pollen of flowers. Honeybees turn nectar into honey to feed their young. Some people **harvest** and sell this honey.

Let's Pollinate!

Bees eat and **collect** pollen. As they fly from one flower to another, they drop some. If the pollen falls on the same kind of flower that it came from, new seeds can be made. This is known as pollination.

Many fruits and vegetables need pollination in order to grow. When new seeds are made, new flowers can grow. Farmers often bring bees to their farms on purpose. By doing this, they can make sure their crops are pollinated and **reproduce**.

While there are other ways for pollination to happen, many plants **depend** on bees to get the job done. A lot of these plants produce foods, such as fruits and vegetables. If bees disappeared, so would many **sources** of food for humans and other animals.

Bees in Trouble

Honeybees are often good at finding their way around. However, sometimes they lose this skill and can't find their way back to their hive. No one knows exactly why this happens. If there aren't enough healthy honeybees in a hive, the **colony** can die off.

In the United States, nearly 100 crops depend on bees for pollination. Many are plants we use for food. If too many bee colonies die, these crops—and the people and animals who eat them—would be in big trouble.

Help the Bees!

Bees might seem like pests when they're flying around you, but they're important parts of nature. Starting a garden is one way you can help bees. Plant flowers native to your area. Your garden will give bees food—and a safe place to live!

GLOSSARY

collect: to get something from a place

colony: a group of animals living and working together

depend: to need something

exist: to be or live

harvest: to bring in a crop

insect: a small, often winged, animal with six legs and three main body parts

reproduce: when a plant or animal creates another just like itself

source: a supplier of something

FOR MORE INFORMATION

BOOKS

Hansen, Grace. *Help the Honey Bees.* Minneapolis, MN: Abdo Kids Jumbo, 2019.

Raatma, Lucia. *A Colony of Bees.* North Mankato, MN: Pebble, 2020.

Vink, Amanda. *Bees Up Close.* New York, NY: PowerKids Press, 2020.

WEBSITES

Bees and Wasps

www.dkfindout.com/us/animals-and-nature/insects/bees-and-wasps/

Learn more about bees and their close relatives, wasps, on this page.

Bee Facts for Kids

www.pestworldforkids.org/pest-guide/bees/

Find fun facts about several kinds of bees on this website.

Honeybee

kids.nationalgeographic.com/animals/invertebrates/facts/honeybee

Check out a map of where honeybees live and read more about these insects here.

INDEX